Editor Karen Barker
Language Consultant Betty Root
Natural History Consultant Dr Gerald Legg

Carolyn Scrace is a graduate of Brighton College of Art, specialising in design and illustration. She has worked in animation, advertising and children's fiction and non-fiction. She is a major contributor to the popular *Worldwise* series and *The X-ray Picture Book* series, particularly **Amazing Animals**, **Your Body** and **Dinosaurs**.

Betty Root was the Director of the Reading and Language Information Centre at the University of Reading for over twenty years. She has worked on numerous children's books, both fiction and non-fiction, and has also held the position of Smarties Book Award Judge.

Dr Gerald Legg holds a doctorate in zoology from Manchester University. His current position is biologist at the Booth Museum of Natural History in Brighton.

David Salariya was born in Dundee, Scotland, where he studied illustration and printmaking, concentrating on book design in his post-graduate year. He has designed and created many new series of children's books for publishers in the U.K. and overseas.

Printed in Belgium.

An SBC Book conceived, edited and designed by
The Salariya Book Company
25 Marlborough Place, Brighton BN1 1UB

A CIP catalogue record for this book is available from the British Library

ISBN 0 7496 3148 1

First published in Great Britain in 1999 by
Franklin Watts
96 Leonard Street
London
EC2A 4RH

Franklin Watts Australia
14 Mars Road
Lane Cove
NSW 2066

lifecycles

The Journey of a Swallow

Written and Illustrated by Carolyn Scrace

Created & Designed by David Salariya

W
FRANKLIN WATTS
NEW YORK • LONDON • SYDNEY

Swallows are small birds.
In autumn, when the weather gets cold, there is less food for them to eat.
Then swallows fly south for thousands of kilometres (see map on page 26) to find warm weather and food.
In the spring they fly back north again to make a nest and breed.

These long journeys are called *migrations*.
In this book you can follow the amazing migration of a swallow.

Swallows have long
thin wings and a large tail.
These help them to fly
great distances.
Swallows find their way
by looking at the position
of the sun and the stars
in the sky.

Swallows use the wind to help them fly.

The moving air lifts and carries them. This means they do not have to flap their wings all the time when they are flying.

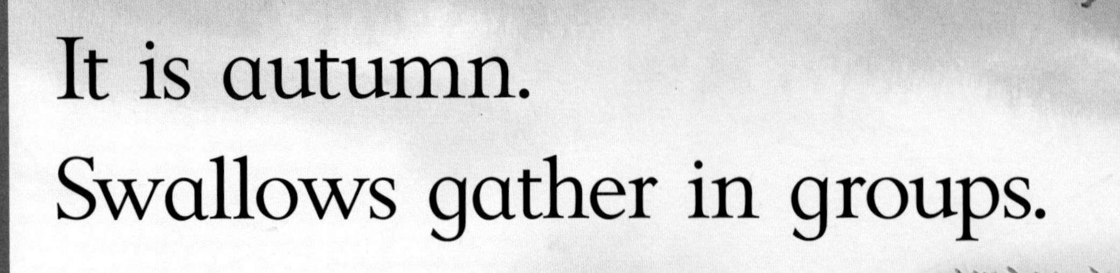

It is autumn.
Swallows gather in groups.

As the weather gets colder,
the birds start their journey south.

Some swallows fly over 10,000 kilometres, from northern Europe to South Africa, or from North America and Canada to Brazil in South America.

Swallows can eat, drink and even sleep while they are in the air. They eat flying insects by opening their beaks very wide and scooping them up.

Some swallows fly over the Sahara Desert in Africa.

The swallows arrive
in the warm south.
Here the air is full of insects.
The birds spend the
winter feeding.

This swallow is flying
low over water
to fill its beak
and drink.

In early February,
swallows start the journey north
to get back home.
Each year they return
to the same place to build a nest.

Swallows like to build their nests in barns or sheds.

Swallows use
mud and straw
to build a nest.
They shape the nest
like a cup, and line it
with feathers and hair.

Nests are often built
on a beam or ledge.

The female swallow lays between 3 and 6 eggs. Both parents take turns to look after the eggs. They sit on the eggs to keep them warm.

It takes 2 weeks for the swallow chicks to grow inside the egg.

A swallow chick uses its beak to break open the eggshell. Both parents catch insects for their chicks to eat.

The hungry chicks are very noisy. Their beaks are bright yellow so that their parents can see them easily.

In 3 weeks, the chicks have grown feathers. Then they are ready to leave the nest.

In the autumn
the young swallows
will start
their own
migration.

ARCTIC OCEAN

NORTH AMERICA

ATLANTIC OCEAN

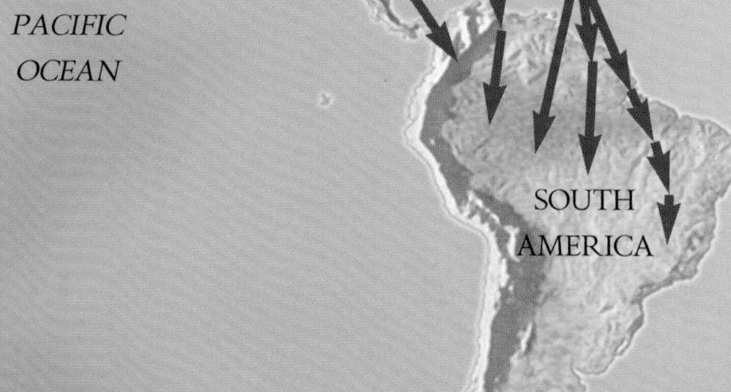

ANTARCTICA

Migration map

Key

migration routes taken by swallows

migration routes taken by arctic terns

migration routes taken by fork-tailed swifts

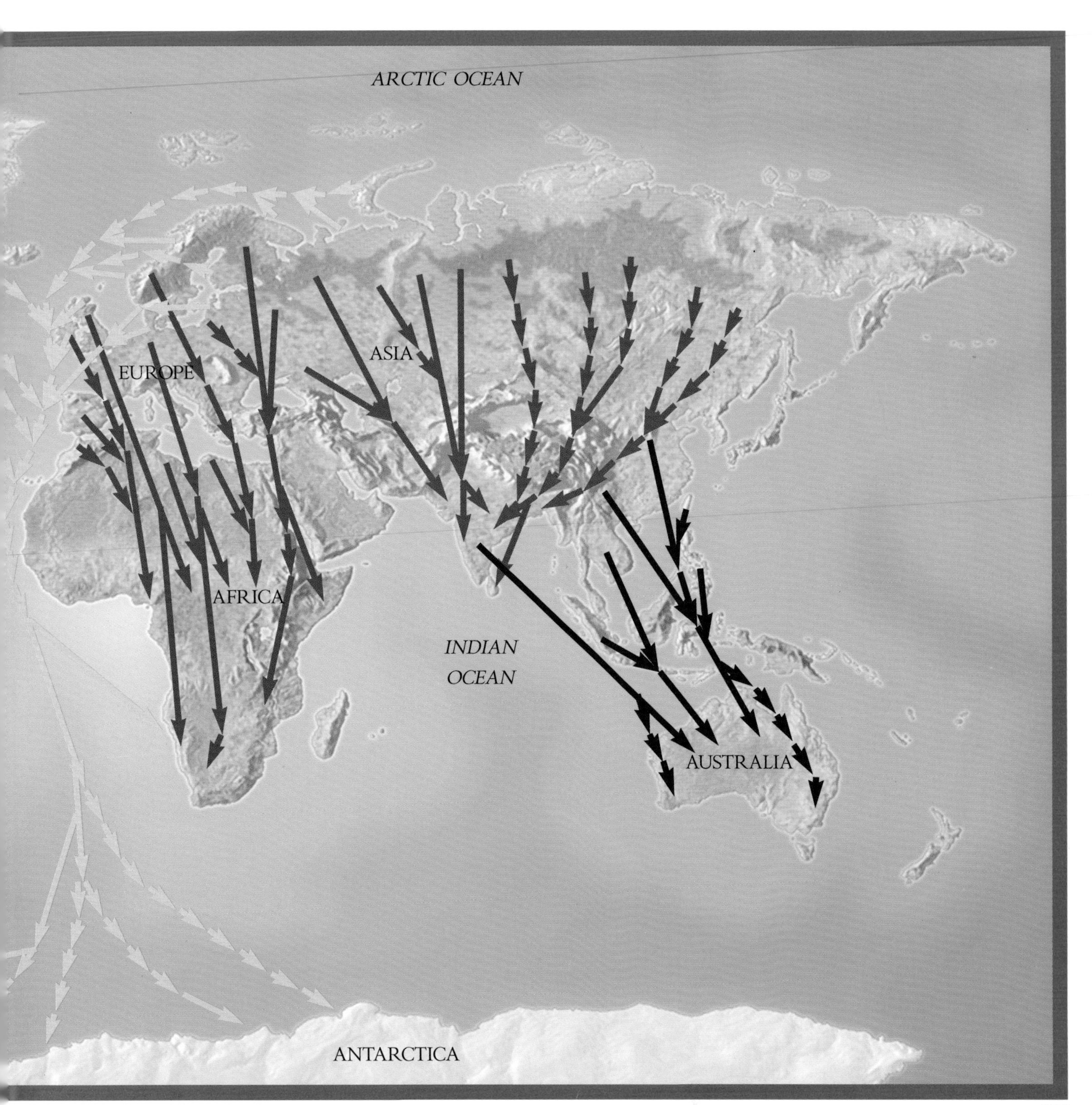
ARCTIC OCEAN
EUROPE
ASIA
AFRICA
INDIAN OCEAN
AUSTRALIA
ANTARCTICA

Swallow words

Beak
A bird's horny mouth.

Breed
To mate and raise a family.

Chick
A baby bird.

Egg
Contains the growing chick before it breaks out.

Feathers
The soft, light coat of birds which helps them to fly. Feathers also keep birds warm and are often colourful.

Insect
A small creature with six legs and a body made up of three parts.

Migration
The long journey made by some animals to find a warmer place to live with plenty of food.

Nest
A hollow place built and used by a bird as a home.

Position
The place where something is. For example, where the sun and stars are in the sky.

Shell
The hard covering of an egg. It keeps the chick inside safe.

Wings
The parts of a bird's body which they use to fly.

Index